Twenty Years of
Conference Room Acoustics
By: Jay Perdue

Twenty Years of Conference Room Acoustics
By: Jay Perdue

About the Author:

Jay Perdue has been involved in conference room acoustics for over twenty years and he holds several patents in architectural acoustics (more patents than anyone else in actual production of products). Jay's passion for learning and understanding in the office conferencing environment, as well as clarity of the spoken word, has led to a strong conviction in the value of acoustics. Jay believes in the science of acoustics, and with several years of working in the field, he has analyzed some tried and true ways to create office, church, and school acoustics (and he explains WHY these methods are so effective). With this book, Jay hopes to give office managers, sales managers and other decision-makers the knowledge on acoustics and the confidence to make successful acoustical choices.

3

Acknowledgments-

"A special thank you goes out to Jeff Phillips, the Harman representative for Sound Marketing Representatives in the Chicago area, for not only making it known to me for the need for such a book as this, but for helping me find resources and even doing some editing and contributions. Thank you so much Jeff for ALL your help!"

I would also like to acknowledge my friends Jim and Chris Stembel at Imagination Pro Media, and their clients, Polywood®, for the beautiful picture and room on the cover. It is their design, not mine, and utilizing Perdue Acoustics products, including printed fabric on the face of several panels...great job Imagination Pro Media!

Also to Angie Allen. Angie came to us as assistant manager and has helped us so much. She took over as manager in tough times and has handled it all with great skill and style, rising to the occasion at every turn. Thanks Angie. If I could I would list every single employee of Perdue Acoustics and give them accolades here. I tell them all the time, Perdue Acoustics is built on my many patents in architectural acoustics under girded by their infinite creative ideas as to how to tell the story of these superior products, manufacture these superior products, install and ship these superior products and everything in between.

Thanks to EVERYONE!!!

Twenty Years of
Conference Room Acoustics
By: Jay Perdue

ISBN NUMBER: 978-1511874939
ISBN-10:1511874937

Twenty Years of Conference Room Acoustics

Jay Perdue - 2015

Introduction

With all the information there is out there on Conference room, Tele Conference Room, and Telepresence Room Acoustics, why would I tackle a booklet on the same subject? First of all, a lot of people have come to depend on me to wade through all the acoustical gobbledygook out there and bring some common sense to it all. Secondly, after reading a LOT of those papers and pamphlets I realized there is a LOT of gobbledygook written on this subject of acoustics as well. I have found statements, written by multimillion dollar, publicly traded telecommunications companies that obviously hired some PHDs to consult on the subject, that make statements so untrue that anyone with two ears knows those statements cannot possibly be true! Impossible?

Exaggeration? Read on and I will bring some quotes out and debunk them for you in a way that you will know, that when all is said and done, common sense and practicality still rules in the science of Architectural Acoustics if you will just let it!

Jay Perdue

Acoustics in the Business Environment
Foreword by Jeff Phillips

Hearing Well in Meetings

Hearing well, when you're in a conference room meeting, is not the same as hearing well when you're joining that meeting via telephone or video conference. It is very common for participants in the room to hear and understand adequately, while the participants joining at a distance struggle to understand, or worse yet, lose attention altogether! Think of the time and money lost when distant members of your team can't accurately comprehend what is being conveyed in a meeting. How does this happen, and how do we fix it?

If you're not in the room, you miss out...

For over 100 years, the telephone has been a brilliant way for humans to communicate quite effectively at a distance. Talkers holding handsets are effectively inches from each other's' ears, where the room acoustics don't have much of an impact. Then 50 years ago came the invention of the speakerphone, and the decline of telephone comprehension began! Imagine you're in a room with your work team. Your brain uses many audible and visual clues to keep you focused on who is speaking, and what is important. You may hear the air conditioner kick on, but your mind quickly filters it out, and it doesn't distract you from the leader's speech.
Side conversations may happen around you, but your ears help your brain place those conversations over THERE, not over HERE, where the leader is speaking. Plus, you can SEE that the leader is still speaking, and follow their lip movements, so you keep your attention where it belongs.

Now, imagine you're listening via telephone. Every sound from the conference is conveyed via one little loudspeaker that you're holding up to your ear. Gone are the audible and visual clues that help you keep your focus on the team leader. The air conditioner becomes a big distraction because your mind can't filter it out. Worse yet, every sound is bouncing all around the meeting room, making it sound "echoey, hollow, mushy, muffled…"

Electronics can take you part of the way, but…

Distance collaboration would not be possible without modern Digital Signal Processing (DSP), which is critical for telephone or video conference meetings. Acoustic Echo Cancellation (AEC), in particular, is crucial to keep far-side audio signals from feeding back from the far-side via microphones in your room. Automatic microphone mixing helps keep "send" levels within a workable range, and a host of other audio fine-tunings are possible. Modern DSP with AEC is an important component, but it can never fix the one thing that often harms your conferences the most…

Not just a good idea, it's The Law! The Law of Physics!!

Optimizing your conference room for tele/video conferencing requires taming the room acoustically. Acoustic treatment usually involves adding absorptive panels to the room, specifically placed to trap excess reflected sound, minimize distraction from noise sources, and help distant participants concentrate on important meeting content.

Often, conference rooms are built to look beautiful and impressive, but are almost useless because they sound awful. Is your conference room awful? A phone call from your room to someone trained in acoustics can place it on the scale from "acoustic treatment needed" to "efficient meeting place" TODAY!

How do we make it better?

For starters, reject any proposal for a tele/video conference system that does not address room acoustics. Rely on AV Integrators who have learned the science of acoustics and who can artfully craft acoustic treatments that will optimize your conference room for comprehension, both within the room, and for distance collaborators.

What is the risk? What is the payoff?

Imagine, in a reverberant room, trying to distinguish the phrase, "We can't do it," from "We can do it." What is the cost of having half the people in the room, and those joining electronically, mishear an important directive? What is the cost to the company, when half the people from the meeting, have a meeting AFTER the meeting, to make sure they understand their marching orders? I'm reminded of the Monty Python skit, where people are straining to hear the words of Jesus, from his sermon on the mount: "I think he said, 'Blessed are the cheese makers...'"

Often, for less than the cost of your speaker phone, you can add acoustic absorption and make sure your meetings are heard crisply, clearly, and with every word understood! The return on your acoustic investment could be measured in days or weeks. Read this book, then act. Be heard and understood today!   *Jeff Phillips*

Chapter I: The Importance of Acoustics

I have been in the acoustics industry for over 25 years. During this time, I have experienced firsthand the frustrations of everyone from the receptionist to the CEO's office when it comes to clear communication. I have also had the opportunity to be responsible for fixing these problems and issues and have received accolades and remarks like "why in the world did we let this go on for so long!" I saw, felt, and heard sound problems from many different perspectives, both in the same room, from room to room, as well as from the room to outside, all contributing to poor communication within a space.

I'm not talking about an individual's inability to either clearly enunciate or clearly present their point, and I'm not talking about an individual's inability to hear these clearly enunciated and presented points. I'm talking about the rooms' ability to garble what has been communicated to the point it was not received by the listener with any chance of clearly UNDERSTANDING what was heard!

Though speakers, microphones, and other communications gear make sure that people can hear the information being given, the acoustics help to ensure people will understand it. The proper acoustics in a project help fine tune a room, create intelligibility and understanding, and make it possible to comfortably enjoy the effort put in by any sound equipment.

It's important to understand acoustics; without the application of acoustics a room can feel too loud, full of echo, impossible to understand what was heard, or even too dead for the overall purpose of the room. It is extremely important to include acoustics in the sound consideration of any business environment. In fact, if a sound design company does not mention or address acoustics, but rather only talks about amplifiers, speakers, microphones, DSP and AEC equipment (more on this later) etc., don't walk away... RUN!

Sound waves have specific "wave lengths." A 10,000 Hz frequency (the range of high sounding cymbals or the sound of an "S" or "T" in diction) has a wavelength of about an inch, a 1,000 Hz frequency (about the pitch of a police siren) has a wavelength of about a foot, and a 100 Hz frequency (just above the range of a kick drum) has a wavelength of about 10 feet. When these sound waves bounce off a reflective wall they can be in perfect harmony, or they can reflect off the wall and completely cancel the frequency (or one of many problems in between, depending on how reflective the surface is).

The problem with these sound reflections is amplifiers and speakers that were designed just fractions of a decibel from being perfectly flat from 50 Hz to 20,000 Hz are put into the room and become plus or minus 100 times that because of the lack of acoustic treatment in the room.

I recently visited the Nashville showroom of one of the best-known home theatre sound companies in the country. I noticed something that I'm not sure most people would notice; all of their demonstration listening rooms were FIVE-SIDED ROOMS. Will they ever sell a system that will go into a five-sided room? Probably not. They just know something that most people don't know; parallel walls create echoes, and parallel walls don't exist in five-sided rooms. The "flat" speakers stay sounding flat. The problem starts when you get the speakers home into your four-sided room and the automatic button to adjust the system to fit the room doesn't work.

You just cannot fix a room that has bad acoustics with equalization. But that's not even the worst part; if the acoustics are bad enough, you may be able to hear the sound system or teleconference system, but you won't be able to understand it. It's not about hearing, it's about UNDERSTANDING. If the rooms where people are being taught, trained, or any discussion is taking place, are excessive in echo and reverberation, the people involved will be able to hear something being spoken, but they will not be able to understand what is being said. Unfortunately, most of the time people use the two words hearing and understanding interchangeably, and they are not at all the same thing.

I remember when we were asked to take a look at the Lubbock Civic Auditorium in Lubbock, TX. The auditorium was a nice, large facility, but all the walls were reflective. The audience was able to hear the plays and musicals, but they were not able to understand what was being said. The facility had to offer headsets to guests to rent in order for the performances to be understood.

Because the facility was used by so many musical production teams, most performances came with their own sound system.

Each sound system ran into the same problems of understanding. This facility provided the best proof that the sound systems were not the problem; the room acoustics needed to be addressed!

The design and installation of the acoustic treatment was to control echoes and excessive reverberation. The very first performance after the acoustic installation was a musical. People came early to get the headsets they normally needed in order to understand the performance. The Lubbock City Manager called to tell us about the "miracle" that occurred that night; every single one of the headsets was turned in at intermission! Every one!

In this facility, where the audience had fought to understand what was being said from the stage for years, the performance had become clearly understood, and it has stayed that way performance after performance, different sound system after different sound system. That's good acoustics!

What happens in a big way happens in a small way. Good fiscal policy for a company is the same as good fiscal policy for a family, just on a smaller scale. Good acoustics making it possible for thousands to understand what's being said in a large auditorium setting translates in the same way to good acoustics making it possible for fewer people being able to understand what's being said in a smaller room... Just on a smaller scale.

Chapter 2: NRC, STC and ICC

Sound Transmission Class (STC) refers to acoustic energy transferring between spaces (how much sound gets out of the room to adjoining rooms) and Noise Reduction Coefficient (NRC) refers to acoustics within rooms (how much and how long the sound bounces around within a space). STC and NRC refer to two completely different worlds of acoustics, and they require two completely different lines of products to treat each one.

Most soft and fuzzy interior acoustical products are targeted at "Noise Reduction Coefficient" (NRC) problems. When a room is echoing or excessively reverberant, it is considered loud, boomy, indistinct, or distorted. In fact, most professional acousticians consider a reverberation time of more than two seconds in large rooms (over 1,000 seats) and a reverberation time of more than one second in conference rooms to be excessive, and it is at this point that diction of the voice and overall clarity are lost. Therefore, even when maximum reverberation is desired it should not exceed two seconds. A simple test can be given to indicate the reverberation time of a room; shout or clap very loudly and see how long it takes for the sound to go away.

There are specific formulas that tell how much of a certain absorptive material is needed to reduce the reverberation in a room to a specified reverberation time. Clarity, intelligibility, and the ability to understand what's being said are lost proportionally as reverberation time rises in any case. There is no such thing as a highly reverberant room that retains clarity.

While reverberation times are limited, it is the job of a good acoustical treatment to eliminate all echoes as well. We will discuss the difference between echo and reverberation and how they affect intelligibility later on. Adding acoustical treatment is the best way to eliminate echo and negate reverberation in a room, restoring clarity and understanding.

Sound Transmission Class problems are altogether different. STC simply refers to how much sound is transmitted from one room or area to the next. This problem is found throughout the working environment.

These sound transmission problems cannot be treated using acoustical wall panels within a room. In an STC situation, cinder block walls filled with sand or double studded sheet rock walls with insulation woven in between become more important in initial construction than anything that can be done after the fact.

As I travel around the country teaching acoustics to Architects and Sound Design Professionals, I've found an illustration to be very helpful here: Imagine, for a moment, you are picking up a teenager at high school. Most kids are driving. As you sit in your modern vehicle with all the windows rolled up tight, you hear them yelling their goodbyes, honking at each other, revving their engines, etc., but it's not really all that loud inside the car. Now, in your imagination, reach over and roll down the window just a half of an inch. Wow! The outside noise comes rolling in! Everything else stayed the same. The same insulation, the same sheet metal, the same glass all around you. All you did was just roll one window, any one window, down a half an inch and all the outside commotion came roaring into your car. That's STC and how it works, or doesn't work! In STC, what's around you is only as good as its weakest 'link'.

Sound takes the path of least resistance, so even the best built wall can have sound transmission problems if the door is not sealed correctly or the room has a drop ceiling and the wall does not go all the way to the roof deck (in this case, sound from one room goes up through the thin ceiling tiles over the wall and down into adjoining rooms). However, there are some helpful hints if the facility is already well past the construction phase.

The cheapest and best fix is to blow insulation into the ceiling as thickly as possible, or add blanket-type insulation tightly packed, eliminating voids. Another way to fix STC issues is to carry the walls up to the building's roof deck. After construction, this can be hard to do in some cases, but it may still be possible. It is also very important to address air vents and ducts. Special duct silencers are available, and board insulation can be used to line the last 2' to 3' inside of the ducts before the air enters or exits the room. Weather stripping around doors and windows, as though they are doors to the outside elements, also helps to seal the room and further eliminate STC problems.

Another issue in STC can be the HVAC system. Once I was called into a school to help with an STC problem. It seemed the school counsellor was having problems with people in the reception area hearing everything that was said in her office. After they took me weaving down this hall and that we finally arrived at her door. I quickly realized that her office, even though a long way from the waiting room door where our journey had begun, was right next to the waiting room!

After further inspection, I discovered the return air vent of both rooms were just two feet apart and connected. Have you ever talked into a pipe to a friend at the other end when you were a child? Yep, same principle! Silencers for sections of HVAC duct are available and planning them further apart with a few turns in them would be nice too! All these helpful hints should head you and your building superintendent in the right direction in either getting situations fixed or knowing where to go and who to ask to

Jay Perdue - 2015

make things better with STC. Here is a simple guide to get you started thinking and doing in the right.

If you're having a problem from street traffic through windows, then your windows do not have a high enough STC rating. Replace the windows with higher STC rated windows, cover the windows with drywall over insulation, cover the windows with acoustic panels, or cover the windows with heavy curtains.

If you're having a problem with noise from a kitchen or bathroom, double the drywall and include a sound barrier material to help reduce the sound, or leave a gap and build another wall with sound barrier material.

If you're having a problem with an adjacent mechanical, copy, work room, break room, or server room, be sure the wall goes all the way up and is attached to the ceiling. If it's not, either continue it on up and seal it to the ceiling deck or blow in or lay out tight insulation above both rooms. You can add acoustical panels to one or both rooms, but remember, an STC fix is only as good as its weakest link, so the entire wall must be treated to be effective. Beyond this, go back to the recommendations of the wet wall and just beef up the wall itself.

If your hearing sound through a door, the first thing to do is weather strip the door so it is air tight. Remember the story of the car after school? If this isn't enough, about the only thing you can do is replace the door with an STC rated door.

The Impact Insulation Class (IIC) is a rating similar to STC but is specific to flooring and ceilings. IIC rating measures the resistance to the transmission of impact noise such as basketballs, chairs dragging, and dropped items. The IIC rating represents the amount of sound energy required to transfer sound through a structure. An IIC rating of 40 would require more than 40 decibels of sound energy to travel through a structure. If this applies to your situation, consult a flooring expert to make right choices here.

Chapter 3: Ambient Noise Levels

Human speech on average is at an amplitude between 60-70 dBA SPL (deciBels of Sound Pressure Level), so it is important to measure this range in any environment where human speech is the primary source of audio. For sound to be intelligible, the amplitude of the intended source must be at least 20 to 30 dBA higher than that of the background and ambient sounds, and in this case the voice must be this much louder than the background noise. Therefore, if human speech is 60 to 70 dBA on average, the background sounds should be lower than 40 to 50 dBA. This is not a specific requirement for teleconference/telecommunication rooms but a fact related to the quality of human hearing and speech intelligibility. The greater the amplitude of the background noise, the harder for the human ear to discern the difference between it and the human voice and to hear accurately. Of course, sounds which are 20 to 30 dBA lower than the human voice can still be heard, however they are considered unintelligible and the human brain will tend to ignore these sounds.

You can buy a dB meter for less than $50 at Radio Shack or download an app for testing sound pressure levels from the App Store. This meter simply tells you how loud the ambient noise is in the room. That means "how loud the room from HVAC is vents and such when nothing is going on in the room." No matter how loud the ambient noise is in the room you have to talk over it, project speech from the other side of the call over it, or turn the volume up from a video or sound recording over it for it to be heard.

You can't have a room at 60 dB ambient noise and talk at 61.dB and be heard, it doesn't work that way. You need to be at least 20dB over the ambient noise to be heard clearly and 30dB would be better. Since most people speak at about 66dB that makes 46dB of ambient noise a maximum and 36 dB preferred. What if you have a CEO or other prominent figure in the company that speaks softly? Then ambient room noise becomes an even greater problem and you better strive hard for the lower number.

There are two weighting scales on a dB meter. The background noise level should be less than 36 dBA and 54 dBC for an office, conference room, or teleconference room. This level is an ideal threshold for background noise in most audio-video communication environments. However, sound levels are usually higher in a commercial office environment. Because of this, Try to lower the background noise to at least 45 dBA and 60 dBA.

The Teleconference systems will continue to operate properly with background noise at even higher levels, however; the sound quality begins to suffer. Background noise above 45 dBA begins to compete with the intelligibility of human speech, and these noises become distracting. Various systems may be able to filter out SOME of the background noise for those at the other end of the call, but the whole communication will be degraded at both ends.

In some cases, when background noise exceeds these limits, gating or sound-suppression effects may occur in various telecommunications systems. These effects happen when the system can no longer distinguish between background noise and what's being said, causing the system to squash most sound because it identifies that sound as background noise.

The most common causes of background noise are: Noise from HVAC air movement, Noise from HVAC machinery or other machinery, external sounds such as street traffic, or adjacent room noise (server rooms, break rooms, bathrooms, or kitchens).

First, figure out where the sound is coming from. Your ears will tell you this as you move silently around the room listening. If the vent itself seems to be making a lot of noise, replace it with a vent that allows less restricted air flow (you may have to separate the ducts into two to increase volume and decrease velocity).

It may be easier to just increase the size of the duct itself as well as the size of the diffuser. Sometimes sweeps are better than 90° bends when it comes to ducting. If the sound seems to be coming from beyond the initial ducting and diffuser, you may need to line the duct with sound dampening material, or add special sound trapping duct work.

Air handlers are found all over modern office spaces. We are used to HVAC units being in one place and pushing air from that one location... More like what we see in homes. But air handlers can be in line to boost air volume and velocity almost anywhere. It might be relatively inexpensive to relocate an air handler further away from your conference room.

Beyond this, it would be best to have an HVAC company come and look at your situation and prescribe solutions. This is beginning to go beyond the scope and knowledge of the company building superintendent in most cases. A final and better solution might just be to relocate the conference room to a more suitable area away from such distractions and noises. That gets back to plain ol' practical!

More of a problem in older buildings, but still worth mentioning here, is ambient noise coming from lighting fixtures. This can be remedied by as simple a task as changing out light fixtures to more modern ones or replacing an old lighting ballast in an old lighting fixture that hums continually.

Other ambient noise problems refer back to the previous chapter. Ambient noise problems can be continual, like HVAC and server room noise, or they can be intermittent like break room noise. In terms of the latter, and where practicality dictates, the conference room use may be dictated by simply scheduling meetings around breaks. Folks can't be in two places at once!

This is a short but important chapter. Maybe I should have just tagged it on to the previous one, but if nothing else, it will stand out in importance just because it got its own chapter, because ambient noise can kill everything. I can't stress enough how important it is to start with a quiet room. But let's go on, because absorption can tame a room that has ambient noise at levels less than ideal as well as bring other problems like echoes, early reflections, and overall reverberance under control.

Jay Perdue - 2015

Chapter 4: Echo and Reverberation

Many people think echo and reverberation are the same, but once you learn the difference, you'll be amazed at how it will catch your ear. You'll not only know the difference, you'll hear the difference.

Reverberation is sound that comes back without distinction. That simply means that you clap your hands and the sound excites the room and it comes back to you seemingly from everywhere, but it doesn't sound like a handclap. In fact, it doesn't sound like anything, you're just aware that a sound has happened in the room of some kind... very indistinct.

Where echoes are categorized solely by their separation in time created by distance, reverberation is specifically quantified in length of time (or how long reverberation goes on). The longer the room allows the sound to reverberate, the harder it is to understand what's being said.

A reverberation time (RT-60) is the time it takes for a sound burst to decay 60 dB in a room. Since 40 dB of ambient room noise is common, I like to say, "The time it takes for a sound burst to decay from 100 db. to 40 db." In some cases, 90 dB down to 30 dB is adequate.

For existing facilities, tests can be performed to see exactly what frequencies are reverberating and at what times. For an immediate close "guess," a loud hard clap, starter pistol, or drum whack and a stopwatch will tell you most of what you need to know (or just go with the loud hard clap and count slowly).

Now, I promised in the introduction of this book to debunk some things written on this subject by some really big and heavy hitters. Here goes the first one... I'll start with a quote from their pamphlet, "Echo is defined similarly to reverberation with one important difference: echo occurs only when the reflection of sound reaches the same location as the original source of the sound."

Wow! Have you ever been at a concert in a coliseum and heard a snare drum whack echo in that huge building? Were you sitting on the drums or next to the drummer? Were you hanging from the rafters next to the speakers that delivered most of the sound to the room? So how could you here an echo that ONLY occurs when you're at the same location as from where the sound originates? In fact you have probably heard thousands of echoes in your life that originated from somewhere else. So has everybody else you know.

So why make such a big deal out of this? To understand echoes and their function in destroying intelligibility is probably the single most fundamental building block in creating good acoustical rooms. I've taught the difference between echo and reverberation to hundreds of Architects and hundreds of Sound a Design professionals, it's really easy to understand, so let me break it down for you here...

Where reverberation is sound returning to you WITHOUT definition or clarity, echo is simply sound returning to you WITH definition or clarity. When you clap your hands, an echo will come back to you as a distinct hand clap. The distance of the surface from which the echo comes back determines if the echo comes back quickly or after a space of time. This is simply a function of distance and the speed of sound, or "how long does it take for sound to get from me to that wall and back".

Since speakers from the platform in lecture halls, auditoriums, and Churches are aimed at the back of the room, when there is an echo that returns to the platform, these echoes are often referred to as "slap back echoes, and they are usually long echoes. Echoes from ceiling to floor or between side walls are often called "flutter echoes," simply because the sound seems to "flutter" really fast between these surfaces due to the usually shorter distances. These echoes, when distinct and drastic enough, can make it impossible to understand the sound system, no matter how good the sound system is. In smaller rooms without sound systems the problem is the same, it's just echoing directly from the speaker's voice instead of through a sound system.

Any echo robs a room of clarity, no matter how distant and slow it is or how close and fast the echo is. To illustrate these echoes, I like to call "slap back" or long echoes "whole word echoes". Slow echoes have time to double double everything everything you you say say, while fast echoes can double a certain consonant such as s, t, or d (D-D-D-DID-D-D-D I SAY THAT-T-T-T?). That's why I have coined the phrase for these echoes as "partial word echoes". These terms really help to get across the concept AND the various problems echoes can cause for intelligibility.

People in our business talk about measuring reverberation and RT60 times like it's the definitive cure all, when, in fact, it's just a simple, beginning test to see if problems might exist in acoustics. Make no mistake about it, echo is the bad boy of acoustics and the arch rival to good intelligibility! Echoes can be problematic in both smaller rooms, and larger rooms. The longer sound travels, the more negatively it can impact the listening experience. To successfully eliminate echo from a room, the large, smooth flat areas that allow echo must be dealt with.

Testing for echo is not nearly as simple as testing for reverberation times. There are many intelligibility tests available, my preference is STI. This test must be performed by a special machine by someone trained in using it. Since limited amounts of reverberation are acceptable but NO apparent echo is, you can see how valuable an STI test can be, I would say down right critical for teleconferencing. We'll talk more on this later...

For facility planning, with a study of the proposed interior finishes, any acoustician can closely predict the RT-60, and look at the overall shape to predict areas of echo production. Because of their very unique intelligibility-robbing qualities and characteristics, we all strive to create rooms of various reverb times but without any noticeable echo whatsoever.

Here's a hint to people who run sound boards...Because digital delay is equal to echo and echo is equal to digital delay and echo robs the speaker of clarity, it is important to use controlled reverberation, NOT digital delay, to add life into the room without robbing it of clarity. Very slight amounts may be desirable for vocal performances, but never a good idea for a speaking voice because the echoes will rob the clarity and intelligibility of what is being conveyed.

That being said, in smaller rooms, you can get away with WAY more acoustical problems in a conference room compared to distance conferencing situations. Even though the conference room should be low in reverberation and absent of all apparent echoes, every little acoustical room issue can show up on the other end of telecommunications, robing the conference of intelligibility and clear communication.

Chapter 5: General Absorption Considerations for Business

Every acoustical environment imaginable can be found in business complexes today; from the smallest office to the largest lecture hall. Every room needs to be treated individually, but the acoustics serve the same purpose in all environments; to create intelligibility and understanding.

It is proven that unless the largest rooms have less than two seconds of reverberation and the smallest rooms have less than a second of reverberation, with little or no apparent echo, our minds cannot connect the dots of the sounds we hear to understand what is being said. We all experience listening fatigue or can't understand what's being said at various levels of echo and reverberation.

We lose our hearing, usually, from the highest frequencies first. Those frequencies are the diction frequencies of the sounds we make. I feel so sorry for older folks sometimes because they are missing these very important enunciations for them to be able to understand what's being said. They may be hearing sound come at them but they are not hearing enough of the high diction frequencies to be able to understand. This makes adult education an even bigger acoustical challenge. The older we get, the worse it gets, and the more important it becomes to teach in good acoustical environments.

If you are going to be LOUD you have to have shorter reverb times to keep it understandable. Volume is a factor of selecting the right RT-60 for a room. The louder the volume, the shorter the reverb time needs to be. The softer the volume, the longer the reverb time CAN be, but never more than two seconds in the largest of rooms.

Putting thinner acoustical products in an acoustical environment that has low frequency volume present may help have an RT-60 of two seconds from 500 Hz and above, but the low frequencies will still be booming around the room for three, four, maybe even five seconds. Remember the older folks? The frequencies they hear the best are the ones that are bouncing around the room. This is totally unacceptable.

Whatever is put into the room must be controlled. If sound is being put into the room down to 500 Hz, the room must be controlled down to 500 Hz. If sound is being put into the room down to 250 Hz, the room must be controlled down to 250 Hz, and if sound is being put into the room down to 125 Hz, the room must be controlled down to 125 Hz.

The fundamental of the male voice is about 125 Hz and the fundamental of the female voice is about 250 Hz. An NRC number used to test and compare acoustical products only starts at 250 Hz! The NRC numbers published for all acoustical products start where the female voice starts and goes up from there (the absorption values don't even include the frequencies where the male voice is!).

Carpet on padding absorbs sound effectively from about 1000 Hz and above. One-inch thick acoustical absorbers made of mineral wool and fiberglass board absorb sound effectively from about 500 Hz and above. Two-inch and three-inch acoustical absorbers made of mineral wool and fiberglass board absorb sound effectively from about 250 Hz and above. The only thing effective at absorbing sound in the 125 Hz range are super thick manufactured and built-in acoustical absorbers of mineral wool and fiberglass board, including my own patents of the MegaWedge™ and 180° Diffsorbers.

This goes to show that we cannot put some carpet or half-inch thick ceiling tiles on the wall and call it an acoustical treatment, because the truth is we've still left the room booming with low frequency energy. That's what the older listener is having the most trouble with. It garbles the sound for all of us, but for them it garbles their whole audible world!

It's important to note that it does cost more to do an acoustical treatment correctly when low frequencies are involved. It costs money to fill a room with low frequency, and it costs money to control it, so it is important to expect that it will.

Chapter 6: Office Acoustics

Let's face it, a loud, reflective, "lively" office can be a hard place to be productive in. It has been definitively proven that people turn off their brains at various lengths of time when they cannot clearly understand the person on the other end of the line well, or there are so many distractions around them they simply cannot concentrate (the same problem occurs in classrooms, and could be the single most important factor in student misbehavior).

Many of these issues go back to STC problems and noise from HVAC and lighting. However the room itself must be dealt with. As we discussed in the chapter on echo and reverberation, if the office itself is fluttering between usually parallel walls, everything that goes on within the space can be a distraction.

Remember, when it comes to echo "it takes two to tango". If one wall is made of glass, you can still kill an office echo by treating the opposite wall. You might get a reflection off the glass one time, but the sound cannot continue to flutter back and forth if the opposing wall is absorptive.

I go back to my years of experience here treating band practice area small practice rooms. If you are not familiar, often smaller office sized rooms are installed in a band rehearsal area for individual practice. The hope is that they will be isolated enough and of a good enough reverberance and echo control to provide a good environment for individual instrument practice.

Jay Perdue - 2015

Of course, in a school situation one wall usually has a large window to keep an eye on things or the door itself was on a shorter wall with the upper area all glass. This led to some very creative reverberation and echo control. The primary thing was to stop continual flutter echoes. I found that if I treated two adjacent walls, not only did I stop the echoes from fluttering back and forth, but I gave the students a "live" corner and a "dead" corner to play into. They could rehearse notes, when first learning a piece, facing the dead corner and then face the live corner when they wanted to perfect their tone for the final performance (kind of get that reflection back to hear what's coming out of the bell of the horn, if you will).

Now, that may seem like a long rabbit trail for no reason, but if you will apply the same principles to your office, you'll be amazed at what a pleasant environment you can come up with, even if you are dealing with reflective walls that cannot, themselves, be dealt with. If you can't stop all the reflections, at least stop all the ones you can, and especially those that are opposite of each other when at all possible.

If you are dealing with two pieces of glass that are opposite from one another, see if you can do without one or the other, or make opposite "halves" of the glass allowed to be covered with acoustical panels. My daughter-in-law, being around acoustics for so long, called and asked if I would have the Perdue Acoustics team make her two panels to fit over our new granddaughters windows in a corner room on a corner lot. Smart girl!

Now flip the page if you don't want a couple of commercials, but I think the next two statements will really help you.

Custom, soft-edged, acoustic panels can be made to exact dimension by Perdue Acoustics in a huge variety of cloth textures and colors. They press right and tight inside the window frame in seconds. There's no reason to have to deal with these annoying reflections in the office environment.

If cost is an issue, Acoustics in a Box, was created to be the least expensive, most effective fix for offices. It comes standard in five colors and one size fits all, but it can fix a whole complex of offices for just a couple hundred dollars per office.

What are the distractions and lack of productivity costing you?

Chapter 7: Conference Room Acoustics

"Ideal conditions for human speech intelligibility are an RT60 value of 300 to 500 milliseconds for all frequencies between 125 Hz and 4,000 Hz." "A very interesting fact is that the range of human hearing is much broader than the range of human speech; the human voice ranges from only about 500 Hz to 2,000 Hz."

Did I just go crazy? Nope. These two statements are quotes from the same publication on acoustics by the same multimillion dollar publicly traded telecommunications giant! So which one is true? The first one. The second "very interesting fact" is not a fact nor true at all.

Have you ever heard the term A 440? That's just the 'A' just above middle 'C' on the piano. 440 just states that when tuning the piano that A is 440Hz, or 440 cycles per second. One octave below that is A 220 and one octave below that is A 110. Now there are a few mega bass singers that can hit this A but the center frequency of the male voice is considered around 125Hz.

Now, why would I make such a big deal out of this? It goes back to the antiquated system we presently use for judging the value of various acoustical absorbers. If you're okay with selecting absorption products that are only effective down to 500 Hz, you're going to miss the whole REAL lower range of the male voice by almost two octaves and the room will be boomy! In fact, you may be missing everything from Middle 'C' on down in frequency.

Can you imagine disregarding all piano keys beneath Middle 'C' (251.6 Hz)? Remember the older folks are losing their hearing from the upper frequencies slowly down. If you leave the room boomy in the lower frequencies, you just made it hardest for them to understand clearly in the room.

I can agree with the reverb times they give in this first statement as well. 300 to 500 milliseconds is just a fancy way of saying about one-third to one-half of a second in reverb time. But, here's another statement I can't agree with in practicality... "An extreme in either direction - too much reverberation or too little - can be detrimental to speech intelligibility." I have yet to see the room that was too dead to the point it was "detrimental to speech intelligibility."

Again, I've heard rooms that were dead in the upper range but left live and boomy in the lower frequencies and speech intelligibility was not good in them; but it was not the deadness of the upper frequencies that was the problem, it was the liveliness and 'boomyness' of the lower frequencies that was causing the problem.

We're putting 100 times more low frequency energy into rooms than we did 50 years ago and still reading publications and treating rooms with products that are fifty-year-old technology. Even in video playback, what sound system would we have used fifty years ago compared to today? Would your video playback of fifty years ago have a subwoofer?

Treat the modern day conference room with lots and lots of low frequency absorption. Everything that takes place in that conference room, from discussion to video presentation, will be more clearly understood. So, where do we put this absorption?

The first priority is to stop any flutter echoes between parallel walls, especially the upper parallel walls. I pioneered some very creative absorbers to do this and catch a maximum amount of low frequency energy at the same time. Many know that low frequency can best be caught in corners, but they think only of the wall to wall intersection.

The wall to ceiling intersection is a corner too. The Wedge™ Series Diffsorbers, with its sloping thickness, depending on which profile you choose, can be only 1" thick at five feet and 7" thick at the wall to ceiling intersection. This provides great wall coverage and great low frequency control at the same time.

Other papers on acoustics suggest lowering the absorbers out of the wall to ceiling corner closer to the floor to achieve a more "lively" room. This, of course, is in direct contradiction to best practices for achieving a best result in low frequency absorption. If there is a need for absorption on the lower part of the wall, a better practice would be to simply add more absorption panels below the upper ones.

For a minimalistic approach, I wouldn't go thinner than 1" thick absorbers in any case, and then only on the lower wall areas to minimize encroachment on the space. Any areas above head height, I would use a thicker panel whenever possible, even a flat one.

Again, the most financially minimalist approach where great acoustics can be achieved is the use of the Acoustics in a Box patented system. It achieves the same kind of overall absorption characteristics with pseudo thickness up in the wall to ceiling corner for maximum acoustical advantage. Aesthetics would be the tradeoff here in both limited color choice and the fact that you can have 2'X4' units, 4'X2' units, or 2'X4' units. Your choice! LOL!!

The lower four to five feet of any conference room usually takes care of it's own echoes, reflections, and excessive reverberations due to furnishings and occupants, but the ceiling can be a special concern in conference rooms, and it's important to note that there can be both benefits and detriments to a reflective ceiling.

First, the benefit, comes from a closer look at how auditorium acoustics have been done in the last century or so. Everyone has seen the hard reflective ceiling in a modern auditorium with all its multiple Angeles ceiling sections. These ceiling sections are at different angles to reinforce sound to the back section of seats. Using acoustics the attempt and purpose here is to make the back seats as close in volume as possible so there is not a bad seat in the house.

In the same way, a reflective ceiling in a conference room can help to reinforce sound from one end to the other as people are talking and discussing. This is especially useful in larger conference rooms with larger tables. These same reflections can, however, do such a good job at carrying sound from one end of the table to another they can make it hard for automatic switching devises to clearly decree who is talking in teleconferencing situations.

So now I'm starting to get ahead of myself and need to close and lead you in to the next chapter on this subject...

Chapter 8: Teleconference/Telepresence Room Acoustics

"Generally reverberation is more of a concern for participants in the room than for the audio being shared with the other side [of a teleconference meeting]. A very flat sound environment will eliminate the harmonics that give speech its uniqueness and personality."

Another two statements from the pages of a very well respected paper written on teleconference room acoustics. Let me break down the direct contradiction of these two statements because they are not as obvious. Jeff Phillips, the writer of the Foreword to this book and the Harman Professional for the Chicago area, talked me into writing this booklet. In his foreword he directly contradicts this first statement and I couldn't agree with him more.

People that sell electronic gear are awfully proud of what that electronic gear can do. I've heard outlandish and misleading statements made on almost every aspect of the electronic side of our industry. One home theater and home sound system company has a one size fits all button that's going to equalize your room no matter what shape your room is. That same company demonstrates its home theaters in five sided rooms because they know more about acoustics they tell you.

Highly directional loudspeakers are a cure all for rooms with bad acoustics according to others, totally ignoring the fact that what comes out of the sound system is only half the problem. The other half of the problem is sitting out there in the thousand people that make up the audience.

So it goes with telecommunications companies that are going to place their microphones so strategically, discern who is talking so perfectly, and reproduce that sound so amazingly on the other side of the world that the room acoustics are more important to those sitting in the room than they are to the ones on the other side of the world?

Maybe if those on the other side of the world are robots with voice recognition capabilities! Psychoacoustics is the study of how the human ears and brain discern sounds compared to analytical machinery. When we're in the room with our two ears we can detect direction, distance, and even block out unwanted sounds around us to concentrate on an individual sound or voice. None of this can be done by folks not in the room.

Then we take into play the second statement in the same brochure about harmonics, a very TRUE statement. The human voice starts with its fundamental, around 125 Hz. For most men and about 250 Hz for most women. From there on up through the human hearing range a series of overtones are produced from every bodies voice in a pattern so unique it's like a fingerprint. This harmonic fingerprint is how we tell the difference in on another's voice.

Other things, like diction and accent, make up a part of this, of course, but I know my wife's voice from nothing more than a short hum... And so do you! Not my wife, yours! Or your husband or significant other or whatever... You know what I mean...

Anyway, how important is it that we reproduce the EXACT sound of the voice speaking across the country without reverberation and echo present in the room so that participants can discern who is speaking? It's only important if you would like to be able to discern between the most important person speaking from the least important person speaking on any given subject. If you are that CEO, you want people to know that it's you speaking compared to your assistant. I know when I'm in a conference call from my Amarillo plant to my Tennessee plant or vice versa, I want people to know who is talking when *I* say "I think ...," compared to the sales assistant saying "I think..."

Acoustics are an important characteristic of any meeting space regardless of the involvement of audio-video systems. Proper acoustics allow the environment to preserve and deliver sound with clarity and accuracy to the human ear from the desired source. For example, in a concert hall, the orchestra and performers on stage are the desired source of sound, and not the audience. Whether the source is a person, a speaker system, or any other instrument, intelligibility of sound can be preserved or impeded based on the acoustics of the environment.

The acoustic characteristics of a room prior to preparation for Telecommunications/teleconferencing may promote sound that is muffled, reverberant, or "echoey". These undesirable effects are often due to noise from airflow from heating and cooling systems, reverberation, or intermittent exterior noise from outside or adjacent rooms already covered in previous chapters. In extreme cases where acoustic factors are not remediated, the audio-detection algorithms used to facilitate switching in multipoint meetings can be adversely affected, resulting in false switching to a participant who is not speaking, or delay in switching to a participant who is speaking.

The acoustic characteristics of a room require close attention during the room selection and design process. Preservation of spatial audio is critical to maintain a life-like virtual meeting. To achieve this objective special emphasis is laid on acoustic elements such as ambient noise, reverberation, and sound isolation.

Just as in Conference Rooms, the walls of the telecommunications/ teleconference rooms must not echo or reverberate much at all. The lower four or five feet will take care of itself so concentrate your absorption on the upper areas of the room. Folks that install telecommunications/ teleconferencing systems on a regular basis tell me that low frequency build up in the room is much more of a problem that has ever been expressed in any literature they have ever read for this type of room.

For this reason, make sure you get an absorber that is good and thick up in the wall to ceiling intersection corner of the room. Perdue Acoustics'Wedge™ Series and Acoustics in a Box's mounting systems were designed for just this purpose. All the walls in teleconference/telecommunications rooms should be treated, because any wall can reflect.

In churches, monitor loudspeakers reflect off of the back platform wall and now their sound is coming back in the right direction to create feedback and other problems. It's the same in the teleconference room. Any wall can reflect a voice from across the room into a microphone, fooling the switching mechanism and causing phantom switching to that microphone.

It would be nice if everyone's voice was the same volume in a teleconference situation, but they're not. Microphones will still pick up a louder voice from across the room than the participant in front of that microphone if the wall is reflective enough. If you can't treat all walls, due to glass and such at the very least treat one of each opposing walls. In this case we recommend a seating arrangement and orientation so that louder voices are speaking towards absorptive panels rather than reflective surfaces.

Sometimes, conference rooms are designed with a sound-reflective ceiling, to carry the sound of softer voices across the room. THIS IS NOT ADVISED for teleconference/telecommunications rooms. ANY reflections can be problematic in these rooms, so it's best to not take a chance on this. I'm not talking about covering every square inch and making the room anechoic. Every other square inch would be good! LOL!

Really, any large area of wall or ceiling left to chance is just that, taking a chance on bad sound! A chance on phantom switching, a chance on system echo, a chance on gating effects, etc. Don't chance it! The deader the better...to a point...and at ALL frequencies, not just the mid frequencies, the low frequencies are especially problematic.

Chapter 9: Acoustics vs. Aesthetics

For 25 years I've seen aesthetic choices take precedence over acoustical choices in business facilities based solely on "what looks good." This war has been waged for so long it demands a close look.

First of all, whether it looks best for the walls to be smooth and flat or to have angular or rounded shapes on them is an opinion. After years of treatment of flat, smooth walls in business facilities with our chunky, angular MegaWedge™ or half round 180° Series I can tell you as many people like the appearance as those that do not.

What is never in question is whether or not people using the facility can understand better. After acoustic treatment, people can understand the person on the phone better or the lecture, or the meeting speaker.

One choice creates an environment where people can understand what's being said and clear communication is obvious. The other choice creates confusion and confused communication so that clear communication is either hampered to various degrees or altogether impossible!

Flat walls echo. Hard walls reverberate. Echo ruins intelligibility. Excessive reverberation creates an environment where clear communication cannot exist. When did the color and shape of the walls become more important than allowing a business leader to be understood? When did contour of the structure get to be more important than clearly understanding so employees can learn and function? The day we started putting aesthetic choice over acoustical fact; that was the day communication problems started.

The aesthetic opinion has gotten into our choices, and when the wrong choice is made, the end result is continual confusion. Will aesthetic choices change the understanding of what is being communicated? No. Will acoustical choices change the understanding of what is being communicated? Absolutely! Every time!

Pictures of the inside of business and office complexes go out into the public in many ways, but only those inside the facility have to deal with the "unseen" acoustical problems.

When we make choices that make the physical building right on the inside, the message gets inside the employees through clear communication and understanding. Without these right choices, the pictures of our new building might look good, but on the inside it's so acoustically rotten it's practically unusable for its intended purpose!

Jay Perdue - 2015

Chapter 10: Acoustical Choices and Testing

There are five ways to judge acoustical absorbers when deciding the best value.

1. Absorption - Not just the NRC rating number, but absorption at all frequencies including the low frequency 125 Hz number. (Type "A" mounting)

2. Aesthetics - Not just how well the company puts the product together and the color and texture choices they have, but the aesthetic choices they have. Do they just have flat looking absorber panels or can you choose angular ones, round ones, etc., to complement and change the look of your room.

3. I Durability - Impact resistance is just the beginning. Some so-called absorbers will fall completely apart if they ever get wet, and others can be easily vandalized. Consider all the possibilities of durability and make sure your choices are right for your room.

4. Fire Protection - This has become a huge factor in recent years. Don't just check the surface burning and smoke characteristics of the product based on the cloth they're covered in, but also find out what the product does in a fire 'as a unit'. You never want acoustical products that contribute to the fire!

5. Value-When you consider ALL the other four factors and then put a price tag on it, which choice is going to give you the best overall value, not which product is the cheapest!

Absorption: To make wise acoustical choices takes knowledge, and just a little math. Sadly, there are those in acoustics who make a living selling "smoke and mirrors" acoustical products, and to make wise choices you have to know what the numbers mean. Beyond what you can see, it's all about absorption per dollars spent and absorption at what

48

frequency! Low frequency absorption is more expensive than upper frequency absorption, and well worth it!

All acoustical products have an NRC rating. An NRC rating is simply a value per square foot of material of absorption. For a room needing 1,400 sq. ft. Sabin's reduction here's the simple math:

It would take 1,000 sq. ft. of NRC 1.40

It would take 1,400 sq. ft. of NRC 1.00

It would take 2,000 sq. ft. of NRC .70

It would take 2,800 sq. ft. of NRC .50

It would take 4,000 sq. ft. of NRC .35

These are not slight differences! 1,000 sq. ft. of a product with an NRC of .35 is NOT equal to 1,000 sq. ft. of a product with an NRC of 1.4; only one fourth the desired absorption is being provided! Now that you know how to compare these products, you can't be fooled.

It is important to note that the NRC rating averages only four frequencies, 250 Hz, 500 Hz, 1,000 Hz, and 2,000 Hz. Back years ago when this system was developed, tweeters and subwoofers were not even in common use. We are creating 10 times the low frequency energy we used to. For this reason, Perdue Acoustics always posts its low 125 Hz test frequency absorption. It is important to always check the 125 Hz absorption numbers and consider them when purchasing any acoustical product.

Always make sure the acoustical tests that are being compared are fair comparisons. Here are some tricks to watch out for.

All tests should be Type "A" tests for comparison. A Type "A" test is done flat on the floor, which equals flat to the wall. Ceiling tiles are often tested 400 centimetres off the test chamber floor, which gives the illusion of a huge low frequency cavity boost that simply does not relate when fastened directly to a wall.

All tests should be the product only, not tested with something else behind it that you don't know about or intend to use. This would be like a test of wood fibre strand board or ceiling tiles over a frame with six inches of fiberglass underneath. Without the total composite construction, the result will be nowhere near the expected result. Read the fine print!

Check the low frequency absorption numbers. It takes both thickness and mass to absorb low frequency energy. Carpet and drapes have mass but no thickness and carved foam products may have thickness but no mass. The numbers tell the story.

There is a product on the market that is called an acoustical absorber, and it is cheap. We were called in to treat a gymnasium that already had this product fully covering all four walls and the ceiling (and they still had a problem with both echo and reverberation). We were able to place our regular absorbers around the walls at one tenth the coverage and the problem was gone! That's because of the value of the absorbers we installed. One-third the square foot price with only one-fourth the absorption is NOT a good value!

Durability:

For the most part, an acoustical absorbers durability is measured in "compressive resistance" (how the panel reacts to an impact) and "tensile strength" (tear strength).

Many fiberglass manufactures face their cores with a compressed 1/8" layer of fiberglass to make them more durable; when one of these "high impact" absorbers is struck, the 1/8" board can crack a hole in the panel, taking the cloth with it. Our company utilizes a stranded fiberglass mat that is the same type mat that is used in the production of boat bottoms. The reinforcing mat flexes and regains 90% compression immediately, and the other 10% over time. You can break through, but it's much harder to do so. Our panels also have a very high tensile strength of 2,631 pounds per square foot!

Fire Protection:

In the standard fire tunnel test, two things are measured: "flame spread" and "smoke developed." Numbers are then given to the product according to these standard testing procedures. For flame spread, Class A requires 25 or less, while the smoke developed rating is 150 or less for some and 450 for others.

The fact is that fiberglass board is rated barely Class A in most tests at flame spread 25. Every manufacturer I have ever researched uses a Class A cloth to cover their product and claims a Class A product, when the fact is the cloth, fiberglass board, and glue that holds it all together burn like crazy as a unit. I do not know how long it will be before the use of such products could be deemed downright negligent! Our products have a flame spread 10 and smoke developed 95 as a whole unit. Check local fire codes and make the best choice based on fire protection.

Quality Diffusion:

All that can be done with sound is absorb it, diffuse it, and reflect it. Most people understand absorption and reflecting; it is diffusion that is a little harder to grasp. Simply put, diffusion turns echo into reverberation. It scatters echo's distinction into reverberation. This has the effect of giving sound fullness and life without the nuisance of the distinct repetition of echo.

So we take a box...all parallel surfaces...an echo nightmare. We take diffusers and diffuse the walls so the parallel walls cannot echo. Now we have a room of no apparent echo, but it reverberates for 4 seconds. This is precisely why we combine units. We need to add in some absorption to bring down the overall reverberation.

Then we might need ceiling reflection to deflect the sound to the back of the lecture hall. Just the right blend of absorption and diffusion will create a room without echo but with exactly the desired amount of reverberation, and then reflection does it's job to get more sound to the back for more even coverage.

But what does a good job of diffusion and what is a good value when diffusers are needed? Since waveforms come in all sizes, a good diffuser needs to represent as many sizes as possible. I patented diffusers in this arena of acoustics because I used smaller chambers within an overall larger "bump." The idea was to let the little chambers of various sizes diffuse the higher and mid frequencies while the overall larger "bump" of the entire unit diffuses the lower frequencies.

These types of diffuser units seem to work best at diffusing the most frequencies for the money. Beyond that, it just gets back to value and aesthetics because most are durable.

Chapter 11: New Findings in Acoustics

Many things in acoustics are moving forward at a rapid rate, but so much of what we do and use in acoustics is fifty years old (old products, old formulas, old test methods, old education, etc.).

Some recent findings negated the much touted Binary Amplitude Diffusion. This product was designed as an absorber that is supposed to have diffusive characteristics as well. The latest studies have shown that the diffusion that this type of product exhibits is extreme near-field, if any at all. When just a few feet away from the panel, the diffusion goes away completely and acts only as a upper frequency partial reflector. The product was found to be not much better than the less expensive vinyl covered absorbers.

Another recent development in acoustics goes a long way to add additional credibility and supports the findings of one of my own patents, the Wedge™ and MegaWedge™ Diffsorbers. The Wedge™ / MegaWedge™ Serieswas created to maximize absorption on the face of the panel while changing the angle of parallel walls and the way they react to one another. Thorough testing of thick flat absorbers (products most commonly used in rooms) explains that thickness works with some frequencies better, while other frequencies just pass through the uniform thickness of the panel much more easily, creating a non-uniform absorption rate in the low frequencies.

The great thing about the Wedge™ / MegaWedge™ Series is the varying thickness of the Diffsorbers, creating a much smoother absorption in the lower frequencies. Uniform thickness creates non-uniform absorption, and non-uniform thickness creates uniform absorption, which is something we all want.

Over the past few years I have had the privilege of talking with some of the greatest minds in acoustics – sometimes arguing with them, and occasionally being right! Thinking outside the box has its advantages. I see things in acoustics that may seem years away, but we are right in the middle of it – and I still have a few acoustical patents up my sleeve!

One of my new patents is a system that I originally called the Wedge Riser System. Marketing people got a hold of it, and it is now known as "The Black Hole Acoustics Mounting System" sold by Acoustics in a Box (**acousticsinabox.com**). This new product utilizes the same sloping absorber technology as the Wedge™ and MegaWedge™ Diffsorbers, but in a "pseudo thickness" method, resulting in less cost and much smaller shipping volumes. "The Black Hole Acoustics Mounting System" is not nearly as durable as the Wedge™ and MegaWedge™ Diffsorbers, due to the fact that it is a one-inch thick absorber with an angled space behind it, but in offices, conference rooms and board rooms on a budget, where durability is less of a factor, they can bring superior absorption at a smaller price.

Variable Room Acoustics is my most recent patent. This acoustical product is designed to transform from a diffuser with an NRC of .60 into an absorber with an NRC of 3.0!!! These units, when used as the primary source of room control, can make an auditorium perfectly right for a string quartet or vocal madrigal performance, and then transform in less than a second for a speech, only to change again to be perfect for an amplified rock concert! VERY exciting news for the acoustics world!

With this new information, acoustical decisions and designs should be easier, helping you to create an environment where the spoken word is clear. If I can help you further or help you to better understand any of these amazing choices we have to make in acoustics, please do not hesitate to call our offices. I will respond to and help you personally whenever possible. THANK YOU for your time in reading this. It's an honour to me that you would do so!

Jay Perdue

www.ingramcontent.com/pod-product-compliance
Lightning Source LLC
Chambersburg PA
CBHW071001180526
45168CB00003B/1240